BIOCHAR HANDBOOK FOR BEGINNERS

The Complete Beginner's Guide To Making Biochar: Supercharge Your Soil, Grow Healthier Plants, And Boost Your Harvest

Emma Ward

Table of Contents

INTRODUCTION TO BIOCHAR 2

    Definition And Historical Background 2

    Importance Of Biochar In Agriculture, Environmental Sustainability, And Climate Change Mitigation 2

Definition and Historical Background 2

Importance of Biochar in Agriculture, Environmental Sustainability, and Climate Change Mitigation 4

CHAPTER ONE 8

SCIENCE BEHIND BIOCHAR 8

    Composition And Properties Of Biochar 8

    Production Methods: Pyrolysis, Hydrothermal Carbonization, Etc. 8

    Chemical And Physical Characteristics That Make Biochar Beneficial 8

Composition and Properties of Biochar 8

Production Methods: Pyrolysis, Hydrothermal Carbonization, etc. 11

Chemical and Physical Characteristics that Make Biochar Beneficial 13

CHAPTER TWO 15

BENEFITS OF BIOCHAR 15

    Soil Health Improvement: Nutrient Retention, Ph Balance, Water Retention 15

    Climate Change Mitigation: Carbon Sequestration, Reducing Greenhouse Gas Emissions 15

    Enhanced Plant Growth And Crop Yields 15

Remediation Of Contaminated Soils 15
1. Soil Health Improvement 15
2. Climate Change Mitigation 17
3. Enhanced Plant Growth and Crop Yields 18
4. Remediation of Contaminated Soils 19
CHAPTER THREE 20
APPLICATIONS OF BIOCHAR 20
   Agriculture: Soil Amendment, Organic Farming, Permaculture 20
   Environmental Remediation: Water Filtration, Landfill Management 20
   Industrial Applications: Energy Production, Construction Materials 20
1. Agriculture 20
2. Environmental Remediation 22
3. Industrial Applications 23
CHAPTER FOUR 25
BIOCHAR PRODUCTION AND TECHNOLOGY 25
   Small-Scale Vs. Large-Scale Production 25
   Innovations In Biochar Production Technologies 25
   Economic Feasibility And Scalability 25
1. Small-Scale vs. Large-Scale Production 25
2. Innovations in Biochar Production Technologies 28
3. Economic Feasibility and Scalability 29
CHAPTER FIVE 32

INTEGRATING BIOCHAR INTO SUSTAINABLE PRACTICES 32
    Case Studies And Success Stories 32
    Best Practices For Application And Dosage 32
    Regulatory Considerations And Safety Guidelines 32
Case Studies and Success Stories 32
Regulatory Considerations and Safety Guidelines 35
CHAPTER SIX 37
FUTURE DIRECTIONS AND INNOVATIONS 37
    Emerging Research Trends 37
    Potential Challenges And Opportunities 37
    The Role Of Biochar In Future Sustainable Development Goals
    37
Emerging Research Trends 37
Potential Challenges and Opportunities 39
A Pocket History of Biochar 41
Step-by-Step Instructions on Making Biochar for Yourself 42
GARDENING WITH BIOCHAR 49
MAKING CHARCOAL AND BIOCHAR 53
BIOCHAR FOR ENVIRONMENTAL MANAGEMENT 58
Forestry and Land Management: 60
CONCLUSION 70
    Summary Of Key Points 70
    Call To Action: Promoting Biochar Adoption And Research .. 70
Summary of Key Points 70
THE END 76

INTRODUCTION TO BIOCHAR

Definition And Historical Background

Importance Of Biochar In Agriculture, Environmental Sustainability, And Climate Change Mitigation

Definition and Historical Background

Biochar is a type of charcoal produced from organic materials through a process called pyrolysis. Pyrolysis involves heating biomass (such as wood chips, crop residues, or animal manure) in the absence of oxygen. This process breaks down the organic material into biochar, along with bio-oil and syngas.

Historical Background: Biochar has been used historically in agriculture by ancient civilizations such as the Amazonians, who created fertile "terra preta" soils by burying charcoal in the ground. This practice dates back thousands of years and demonstrates biochar's potential to improve soil fertility and productivity.

Importance of Biochar in Agriculture, Environmental Sustainability, and Climate Change Mitigation

Agriculture:

Soil Fertility: Biochar improves soil structure, water retention, and nutrient availability. It acts like a sponge, holding onto water and nutrients, thereby reducing leaching and improving plant resilience to drought.

Microbial Activity: It enhances soil microbial communities, promoting beneficial microbial activities that aid in nutrient cycling and plant health.

Crop Yield: Studies have shown increased crop yields in biochar-amended soils due to improved soil fertility and water holding capacity.

Environmental Sustainability:

Carbon Sequestration: Biochar is a stable form of carbon that can remain in soils for hundreds to thousands of years, effectively sequestering carbon and mitigating greenhouse gas emissions.

Waste Management: It provides a beneficial use for agricultural residues, forestry by-products, and organic waste materials that might otherwise be burned or landfilled, contributing to waste reduction and sustainable resource management.

Climate Change Mitigation:

Reduced Emissions: By sequestering carbon and improving soil health, biochar helps reduce emissions of carbon dioxide (CO_2) and other greenhouse gases associated with conventional agricultural practices.

Mitigation of Soil Degradation: Biochar application can mitigate soil degradation caused by intensive farming practices, erosion, and loss of soil organic matter.

CHAPTER ONE

SCIENCE BEHIND BIOCHAR

Composition And Properties Of Biochar

Production Methods: Pyrolysis, Hydrothermal Carbonization, Etc.

Chemical And Physical Characteristics That Make Biochar Beneficial

Composition and Properties of Biochar

Composition: Biochar is primarily composed of carbon, typically ranging from 60% to 90% by weight. Its composition also includes ash, which varies depending on the feedstock used for production. Minor components may include volatile organic compounds, water, and gases trapped within the structure.

Properties: Biochar exhibits several physical and chemical properties that contribute to its beneficial effects in agriculture and environmental applications:

Porosity: Biochar has a highly porous structure with a large surface area, which allows it to retain water and nutrients in soils. This porosity also provides habitat for soil microorganisms.

Surface Area: The high surface area of biochar enhances its ability to adsorb and retain nutrients, contaminants, and organic compounds in soils, thereby improving soil fertility and reducing nutrient leaching.

Stability: Biochar is resistant to decomposition and can persist in soils for extended periods, contributing to long-term soil carbon storage and climate change mitigation.

Production Methods: Pyrolysis, Hydrothermal Carbonization, etc.

Pyrolysis: This is the most common method for biochar production. It involves heating biomass (e.g., wood chips, agricultural residues) in the absence of oxygen (anaerobic conditions) at temperatures typically ranging from 300°C to 700°C. The process decomposes the biomass into biochar, bio-oil, and syngas.

Hydrothermal Carbonization (HTC): HTC is a wet thermal conversion process that operates at lower temperatures (180°C to 250°C) and higher pressures. Biomass is treated with water under these conditions to produce a solid biochar-like material with different properties compared to pyrolysis biochar.

Gasification: Biomass is partially oxidized with a controlled amount of oxygen or steam at high temperatures (700°C to 1000°C) to produce a synthesis gas (syngas) and biochar.

Each production method influences the properties of biochar, affecting its porosity, surface area, and chemical composition, which in turn impact its effectiveness in agricultural and environmental applications.

Chemical and Physical Characteristics that Make Biochar Beneficial

Nutrient Retention: Biochar's porous structure and high surface area enable it to retain water, nutrients (such as nitrogen, phosphorus, and potassium), and other essential elements in soils. This improves nutrient availability for plants and reduces nutrient runoff.

pH Buffering: Biochar can buffer soil pH, helping to maintain optimal conditions for plant growth and microbial activity.

Adsorption of Contaminants: Biochar has a strong affinity for organic pollutants, heavy metals, and other contaminants in soil and water, effectively reducing their bioavailability and environmental impact.

Microbial Habitat: Biochar provides a stable habitat for soil microorganisms, enhancing soil biodiversity and promoting beneficial microbial activities that contribute to soil health and fertility.

CHAPTER TWO

BENEFITS OF BIOCHAR

Soil Health Improvement: Nutrient Retention, Ph Balance, Water Retention

Climate Change Mitigation: Carbon Sequestration, Reducing Greenhouse Gas Emissions

Enhanced Plant Growth And Crop Yields

Remediation Of Contaminated Soils

1. Soil Health Improvement

Nutrient Retention: Biochar enhances soil fertility by adsorbing and retaining nutrients such as nitrogen, phosphorus, and potassium. This helps reduce nutrient leaching and improves nutrient availability for plants over time.

pH Balance: Biochar acts as a pH buffer, helping to stabilize soil pH levels. This is particularly beneficial in acidic or alkaline soils, where biochar can adjust pH towards a more neutral range, promoting optimal conditions for plant growth.

Water Retention: The porous structure of biochar enables it to retain water and improve soil water holding capacity. This helps plants access water during dry periods, reducing irrigation needs and enhancing drought resistance.

2. Climate Change Mitigation

Carbon Sequestration: Biochar is a stable form of carbon that can remain in soils for hundreds to thousands of years. By burying carbon in agricultural soils, biochar helps mitigate climate change by sequestering carbon dioxide (CO_2) from the atmosphere.

Reducing Greenhouse Gas Emissions: Biochar application can reduce emissions of nitrous oxide (N_2O) and methane (CH_4) from soils, which are potent greenhouse gases associated with conventional agricultural practices.

3. Enhanced Plant Growth and Crop Yields

Improved Soil Fertility: Biochar enhances soil structure and nutrient availability, promoting healthier root growth and nutrient uptake by plants. This contributes to improved plant growth, vigor, and overall crop yields.

Water Efficiency: By improving soil water retention and nutrient availability, biochar helps plants maintain optimal hydration levels and reduces water stress, leading to better growth and productivity.

4. Remediation of Contaminated Soils

Adsorption of Contaminants: Biochar has a high surface area and strong adsorption capacity for heavy metals, organic pollutants, and other contaminants in soils. This can help remediate contaminated sites by reducing the bioavailability and mobility of pollutants, thereby improving soil quality and reducing environmental risks.

CHAPTER THREE

APPLICATIONS OF BIOCHAR

Agriculture: Soil Amendment, Organic Farming, Permaculture

Environmental Remediation: Water Filtration, Landfill Management

Industrial Applications: Energy Production, Construction Materials

1. Agriculture

Soil Amendment: Biochar is widely used as a soil amendment to improve soil health and fertility. It enhances nutrient retention, water holding capacity, and soil structure, thereby promoting healthier plant growth and increased crop yields.

Organic Farming: Biochar is compatible with organic farming practices as it is derived from natural, renewable biomass sources. It helps organic farmers improve soil quality without relying on synthetic fertilizers or chemicals.

Permaculture: In permaculture systems, biochar is integrated into sustainable farming practices to enhance soil biodiversity, promote water conservation, and support resilient agricultural ecosystems.

2. Environmental Remediation

Water Filtration: Biochar's high surface area and adsorption capacity make it effective for water filtration and purification. It can be used in filters to remove contaminants such as heavy metals, organic pollutants, and pathogens from water sources.

Landfill Management: Biochar can be applied in landfill covers or as a soil amendment in landfill reclamation projects. It helps mitigate odors, reduce leachate contamination, and improve soil conditions for re-vegetation and land restoration.

3. Industrial Applications

Energy Production: Biochar can be used as a biomass feedstock for energy production through processes like gasification or combustion. It provides a renewable source of energy while also producing biochar as a byproduct that can be utilized for soil improvement.

Construction Materials: Biochar can be incorporated into construction materials such as concrete and plaster to enhance their thermal and mechanical properties. This application contributes to sustainable building practices by reducing environmental impacts and improving building performance.

CHAPTER FOUR

BIOCHAR PRODUCTION AND TECHNOLOGY

Small-Scale Vs. Large-Scale Production

Innovations In Biochar Production Technologies

Economic Feasibility And Scalability

1. Small-Scale vs. Large-Scale Production

Small-Scale Production:

Methods: Small-scale biochar production typically involves simple technologies such as kilns or retorts. These methods are suitable for local or community-based initiatives where biomass feedstocks are readily available.

Benefits: It allows for decentralized production, providing farmers and small-scale growers with access to biochar for soil improvement and agricultural purposes.

Challenges: Efficiency and consistency of biochar quality can vary due to the manual operation and limited control over process parameters.

Large-Scale Production:

Methods: Large-scale biochar production utilizes advanced

technologies such as continuous pyrolysis systems or rotary kilns. These systems are designed for higher throughput and efficiency.

Benefits: They can produce biochar in larger quantities with better control over process parameters, ensuring consistent quality and properties.

Challenges: Initial investment costs can be significant, and operational complexity requires skilled personnel and infrastructure for handling feedstocks and managing emissions.

2. Innovations in Biochar Production Technologies

Pyrolysis Technologies: Continuous pyrolysis systems have been developed to optimize heat transfer, maximize biochar yield, and minimize emissions. These technologies often incorporate heat recovery and gas cleaning systems to improve energy efficiency and reduce environmental impact.

Hydrothermal Carbonization (HTC): HTC processes have advanced to produce hydrochar, a biochar-like material with different properties than traditional biochar. HTC operates at lower temperatures and pressures, offering potential advantages in processing wet biomass and producing biochar under mild conditions.

Integrated Systems: Innovations focus on integrating biochar production with other processes, such as renewable energy generation (via syngas) or wastewater treatment, to enhance overall resource efficiency and economic viability.

3. Economic Feasibility and Scalability

Economic Feasibility: The economic viability of biochar production depends on factors such as feedstock availability, production scale, energy costs, and market demand for biochar products. Innovations in technology aim to reduce production costs and improve profitability by increasing biochar yield and enhancing process efficiency.

Scalability: Advances in production technologies have enabled scalability from small-scale operations serving local communities to large-scale facilities supplying regional or national markets. Scalability is crucial for widespread adoption of biochar in agriculture, environmental remediation, and industrial applications.

CHAPTER FIVE

INTEGRATING BIOCHAR INTO SUSTAINABLE PRACTICES

Case Studies And Success Stories

Best Practices For Application And Dosage

Regulatory Considerations And Safety Guidelines

Case Studies and Success Stories

Terra Preta in the Amazon: Historical use of biochar by ancient Amazonian civilizations (Terra Preta soils) demonstrates long-term benefits in soil fertility, carbon sequestration, and sustainable agriculture.

Modern Agriculture: Numerous case studies highlight biochar's positive impacts on crop yields, soil health improvement, and water retention in diverse agricultural settings worldwide.

Environmental Remediation: Biochar has been successfully used to remediate contaminated soils and improve water quality in polluted areas, showcasing its effectiveness in environmental sustainability.

Best Practices for Application and Dosage

Soil Preparation: Biochar should be incorporated into soils during initial soil preparation or planting to ensure uniform distribution and maximum contact with roots.

Dosage: Recommended biochar application rates vary based on soil type, crop, and specific objectives. Generally, initial application rates range from 1 to 10% by volume, with higher rates for degraded soils or intensive agriculture.

Mixing and Incorporation: Proper mixing and incorporation of biochar into soil layers help optimize its benefits, ensuring even distribution and minimal surface exposure.

Regulatory Considerations and Safety Guidelines

Regulatory Framework: Regulatory requirements for biochar vary by region and application. Some jurisdictions classify biochar as a soil amendment or fertilizer, while others regulate it under waste management or environmental protection laws.

Safety Guidelines: Biochar should meet quality standards to ensure absence of contaminants (e.g., heavy metals, pathogens) that could impact soil health or human health. Guidelines often recommend testing biochar for pH, nutrient content, and potential contaminants before application.

Environmental Impact: Assessments consider biochar's impact on soil pH, nutrient cycling, and potential leaching of contaminants into groundwater or surface water.

CHAPTER SIX

FUTURE DIRECTIONS AND INNOVATIONS

Emerging Research Trends

Potential Challenges And Opportunities

The Role Of Biochar In Future Sustainable Development Goals

Emerging Research Trends

Advanced Production Technologies: Continued research focuses on enhancing biochar production efficiency, reducing energy consumption, and developing novel pyrolysis and hydrothermal carbonization techniques.

Biochar Applications: Research explores new applications of biochar beyond agriculture, including carbon capture and storage, wastewater treatment, and bioremediation of contaminated sites.

Biochar-Enhanced Materials: Investigation into integrating biochar into composite materials, such as building materials and energy storage devices, to enhance their properties and sustainability.

Potential Challenges and Opportunities

Scaling Production: Scaling up biochar production to meet global demand while maintaining economic feasibility and environmental sustainability remains a challenge.

Public Perception: Educating stakeholders about biochar's benefits, safety, and regulatory compliance is essential to overcome misconceptions and promote wider adoption.

Feedstock Availability: Ensuring sustainable sourcing of biomass feedstocks for biochar production without competing with food or causing land use conflicts is critical for long-term viability.

The Role of Biochar in Future Sustainable Development Goals

Climate Change Mitigation: Biochar's ability to sequester carbon and reduce greenhouse gas emissions aligns with global efforts to mitigate climate change and achieve carbon neutrality.

Soil Health and Food Security: By improving soil fertility, water retention, and nutrient availability, biochar contributes to sustainable agriculture, enhancing food security and resilience to climate variability.

Circular Economy: Biochar supports circular economy principles by converting biomass waste into valuable resources, promoting resource efficiency and reducing environmental impacts.

A Pocket History of Biochar

Biochar has a rich history dating back thousands of years, notably used by ancient civilizations like the Amazonians in creating fertile "terra preta" soils. These soils, enriched with biochar, demonstrated enhanced fertility and carbon sequestration capabilities, offering a sustainable model for agriculture and soil management.

Step-by-Step Instructions on Making Biochar for Yourself

Gather Biomass: Collect dry biomass such as wood chips, crop residues, or yard trimmings. Avoid materials treated with chemicals or paints.

Choose a Container: Select a metal drum or airtight container with a lid. Make sure it's clean and free of residues.

Prepare for Pyrolysis:

Fill the container with biomass loosely packed, leaving some space at the top.

Drill small holes near the bottom for airflow (optional).

Start Pyrolysis:

Place the container on a fireproof surface outdoors or in a well-ventilated area.

Light a small fire underneath and gradually increase the heat.

Cover the container with the lid but leave it slightly ajar to allow gases to escape.

Monitor the Process:

Pyrolysis typically takes several hours. Monitor the temperature and adjust airflow as needed to maintain a slow, steady burn.

Cooling and Storage:

Once pyrolysis is complete (when smoke decreases and turns clear), let the container cool completely.

Store biochar in a dry, covered container to prevent it from absorbing moisture.

Applications for Biochar

Soil Water Retention: Mix biochar into soil to improve water retention and reduce irrigation needs, especially in sandy or loamy soils.

Pest Deterrence: Use biochar as a natural pest deterrent by applying it around plants or mixing it with compost to discourage pests and pathogens.

Compost Enhancement: Incorporate biochar into compost piles to improve nutrient retention and microbial activity, accelerating decomposition and producing richer compost.

Inspiring Examples of Ecosystem Restoration and Improved Forest Management

Forest Soil Restoration: Biochar has been used to restore degraded forest soils, improving soil fertility and enhancing tree growth in areas affected by logging or wildfire.

Agroforestry Systems: Incorporating biochar into agroforestry practices promotes sustainable land management, increases crop resilience, and enhances biodiversity.

Low-Cost Recipes, including Cultured Biochar and Sustainable Potting Soil

Cultured Biochar: Mix biochar with compost tea or liquid organic fertilizers to create a nutrient-rich solution for plants, enhancing soil fertility and plant health.

Sustainable Potting Soil: Blend biochar with coconut coir or peat moss, perlite, and compost to create a lightweight, nutrient-rich potting mix suitable for container gardening.

By exploring these practical applications and examples, biochar emerges as a versatile tool for sustainable agriculture, ecosystem restoration, and innovative gardening practices, offering tangible benefits for soil health, water conservation, and environmental stewardship.

GARDENING WITH BIOCHAR

Gardening with biochar can be highly beneficial for your plants and soil. Biochar is essentially charcoal produced from organic materials like wood, crop residues, or manure, created through a process called pyrolysis (heating in the absence of oxygen). Here are some key benefits and tips for using biochar in gardening:

Benefits of Biochar in Gardening:

Improved Soil Fertility: Biochar enhances soil fertility by improving nutrient retention, especially for essential elements like nitrogen, phosphorus, and potassium.

Enhanced Soil Structure: It helps improve soil structure by promoting better drainage and aeration, which is crucial for root health and water retention.

Increased Microbial Activity: Biochar provides a habitat for beneficial soil microbes, which aid in nutrient cycling and overall soil health.

Carbon Sequestration: It sequesters carbon in the soil for hundreds to thousands of years, contributing to climate change mitigation.

Tips for Gardening with Biochar:

Application: Mix biochar thoroughly into the soil. It's best to charge biochar before application by soaking it in compost tea or nutrient-rich water to enhance its effectiveness.

Amount: Apply biochar at a rate of about 5-10% by volume of your soil. Too much biochar initially can absorb nutrients and hinder plant growth until it's charged with nutrients.

Types of Biochar: Consider the source of biochar (e.g., wood-based, agricultural residues) as it can affect nutrient content and pH of the soil.

Seasonal Application: Apply biochar during the growing season to maximize its benefits, or incorporate it during soil preparation before planting.

Long-Term Benefits: Biochar's effects improve over time as it interacts with soil microbiota and continues to enhance soil structure and fertility.

Complementary Practices: Use biochar in conjunction with organic fertilizers and compost to provide a balanced nutrient supply for plants.

MAKING CHARCOAL AND BIOCHAR

Making charcoal and biochar involves a process called pyrolysis, which is essentially heating organic material in the absence of oxygen. Here's a basic overview of how you can make charcoal and biochar:

Making Charcoal:

Material Selection: Choose hardwoods such as oak, maple, or hickory for making charcoal. Softwoods like pine can also be used but may produce a lower-quality charcoal.

Carbonization Process:

Traditional Method: Stack wood in a mound or pyramid shape, cover it with a layer of soil or clay to limit oxygen supply, and light it from the top. This process slowly carbonizes the wood over several hours to days.

Kiln Method: Use a metal or brick kiln to heat the wood at high temperatures (around 500-600°C) in a controlled environment to produce charcoal.

Cooling and Harvesting: Once the wood is carbonized and cooled, break open the mound or kiln to retrieve the charcoal. It should be blackened and brittle.

Making Biochar:

Material Selection: Biochar can be made from a variety of organic materials such as wood chips, agricultural residues (like corn cobs or rice husks), or even animal manure.

Pyrolysis Process:

Small-Scale Method: Place the organic material in a metal drum or container with a lid that can seal tightly. Heat the container over a fire or in an oven while limiting oxygen intake. This process can take several hours.

Large-Scale Method: Use specially designed biochar kilns or reactors that allow for controlled heating and gas capture, which can be more efficient for larger quantities.

Cooling and Storage: After pyrolysis, let the biochar cool down. It should be porous and blackened. Store biochar in a dry place to prevent it from absorbing moisture.

Considerations:

Safety: Both charcoal and biochar production involve high temperatures and should be done with caution to avoid fire hazards and inhalation of smoke or gases.

Quality: The quality of charcoal and biochar depends on the materials used and the pyrolysis process. Higher temperatures generally produce better-quality charcoal and biochar.

Uses: Charcoal is primarily used for cooking and heating, while biochar is used in agriculture and gardening to improve soil fertility and structure.

BIOCHAR FOR ENVIRONMENTAL MANAGEMENT

Biochar offers several benefits for environmental management, particularly in agriculture, forestry, and waste management. Here are some key ways biochar can be used for environmental purposes:

Soil Improvement:

Enhanced Soil Fertility: Biochar improves nutrient retention in soils, reducing the need for chemical fertilizers and promoting sustainable agriculture.

Carbon Sequestration: By locking carbon into the soil for long periods (potentially centuries), biochar contributes to carbon sequestration, mitigating climate change.

Improved Water Retention: Biochar enhances soil structure, increasing water retention capacity and reducing water runoff and erosion.

Waste Management:

Organic Waste Utilization: Biochar can be made from agricultural residues, forestry waste, and even some types of municipal organic waste, providing a sustainable way to manage these materials.

Reduced Greenhouse Gas Emissions: Converting organic waste into biochar reduces methane emissions from anaerobic decomposition and reduces overall greenhouse gas emissions.

Forestry and Land Management: Wildfire Risk Reduction: Applying biochar to forest soils can improve soil health and reduce the risk of severe wildfires by enhancing moisture retention and nutrient availability.

Soil Remediation: Biochar can be used to remediate contaminated soils by adsorbing heavy metals and toxins, reducing their availability to plants and mitigating environmental pollution.

Energy and Industry:

Renewable Energy Production: The process of producing biochar (pyrolysis) can generate bioenergy (syngas) that can be used for heat and power generation, contributing to renewable energy goals.

Industrial Applications: Biochar has applications in various industrial processes, such as water filtration, where its high surface area and porous structure make it effective in adsorbing contaminants.

Challenges and Considerations:

Application and Integration: Effective use of biochar requires understanding local soil conditions, optimal application rates, and crop-specific benefits.

Long-Term Effects: While biochar offers long-term soil benefits, its effects can vary based on soil type, climate, and management practices.

Economic Viability: Adoption of biochar in large-scale applications may require economic incentives or policies that promote sustainable practices and carbon sequestration.

Biochar Solution

The term "biochar solution" typically refers to the application of biochar as a sustainable and beneficial soil amendment in various environmental and agricultural contexts. Here's a structured approach to understanding and implementing a biochar solution:

Understanding Biochar:

Definition: Biochar is a carbon-rich material produced from organic biomass (such as wood chips, agricultural residues, or manure) through a process called pyrolysis, which involves heating in the absence of oxygen.

Properties: Biochar is porous, stable in the soil for long periods (hundreds to thousands of years), and has a high surface area, which enhances its ability to retain nutrients and water.

Benefits of Biochar:

Soil Fertility: Improves nutrient retention and availability in soils, reducing the need for chemical fertilizers and enhancing crop productivity.

Carbon Sequestration: Sequesters carbon in the soil, mitigating greenhouse gas emissions and contributing to climate change mitigation efforts.

Water Management: Enhances soil structure, promoting better water retention and reducing water runoff and erosion.

Soil Health: Supports beneficial soil microorganisms, improving overall soil health and resilience to environmental stressors.

Implementing a Biochar Solution:

Selection of Biochar: Choose biochar derived from suitable feedstocks and produced under controlled conditions to ensure quality and effectiveness.

Application Rates: Determine appropriate application rates based on soil type, crop requirements, and local conditions. Generally, apply biochar at rates of 5-10% by volume of soil.

Application Methods: Mix biochar thoroughly into the topsoil during planting or incorporate it into compost before application to enhance its nutrient-loading capacity.

Monitoring and Evaluation: Monitor soil conditions, crop growth, and nutrient levels over time to assess the impact of biochar on soil health and productivity.

Integration with Sustainable Practices: Integrate biochar use with other sustainable agricultural practices, such as organic farming methods and efficient water management, to maximize benefits.

Challenges and Considerations:

Cost and Accessibility: Initial costs of biochar production and application may vary, but long-term benefits often outweigh initial investments, especially in terms of soil health and productivity.

Research and Adaptation: Stay updated on research findings and adapt biochar application strategies to local agricultural and environmental conditions for optimal results.

Policy Support: Advocate for policies that promote sustainable soil management practices and support the use of biochar as a climate-smart agricultural solution.

CONCLUSION

Summary Of Key Points

Call To Action: Promoting Biochar Adoption And Research

Summary of Key Points

Biochar Definition and Composition: Biochar is a carbon-rich material produced from biomass through pyrolysis or hydrothermal carbonization, known for its porous structure and ability to improve soil health.

Benefits of Biochar: It enhances soil fertility, water retention, and nutrient availability, mitigates climate change by sequestering carbon, improves plant growth and crop yields, and remediates contaminated soils.

Applications: Biochar is used in agriculture for soil amendment, in environmental remediation for water filtration and landfill management, and in industrial applications such as energy production and construction materials.

Integration into Sustainable Practices: Best practices include proper application and dosage, adherence to safety guidelines and regulatory considerations, and learning from successful case studies globally.

Future Directions and Innovations: Emerging research focuses on advanced production technologies, expanding biochar applications, addressing challenges like scaling production and feedstock availability, and advancing biochar's role in achieving sustainable development goals.

Call to Action: Promoting Biochar Adoption and Research

Biochar offers tangible solutions to global challenges such as climate change, soil degradation, and environmental pollution. To promote its adoption and research:

Educate Stakeholders: Raise awareness about biochar's benefits, safety, and regulatory compliance among farmers, policymakers, industries, and researchers.

Support Research and Innovation: Invest in research to advance biochar production technologies, explore new applications, and address scalability and sustainability challenges.

Policy and Regulatory Support: Develop supportive policies and regulations that facilitate biochar adoption, ensure quality standards, and promote sustainable sourcing of biomass feedstocks.

Collaboration and Knowledge Sharing: Foster partnerships between academia, industry, government agencies, and communities to share knowledge, best practices, and case studies.

By promoting biochar adoption and research, we can harness its potential to contribute significantly to sustainable development goals, enhance agricultural resilience, and mitigate climate change impacts globally.

THE END

www.ingramcontent.com/pod-product-compliance
Lightning Source LLC
Chambersburg PA
CBHW071840210526
45479CB00001B/226